Science Sight Word Readers

Ponds

by Casey Lasko

No part of this publication may be reproduced, stored in a retrieval system, or transmitted in any form or by any means, electronic, mechanical, photocopying, recording, or otherwise, without written permission of the publisher. For information regarding permission, write to Scholastic Inc., Attention: Permissions Department, 557 Broadway, New York, NY 10012.

ISBN 978-0-545-24807-5

Photographs © 2010: cover: Alamy Images/Les Gibbon; back cover top: iStockphoto/Zoran Kolundzija; back cover bottom: Minden Pictures/Wil Meinderts; page 1: iStockphoto/Pavel Mozzhukhin; page 2: ShutterStock, Inc./Stephen Orsillo; page 3: Getty Images/David W. Hamilton; page 4: Getty Images/James Hager; page 5 main: iStockphoto/Bruce MacQueen; page 5 inset: ShutterStock, Inc./James "BO" Insogna; page 6 main: iStockphoto/Hedda Gjerpen; page 6 inset: iStockphoto/Tomasz Zachariasz; page 7 main: iStockphoto/Stefan Gerzoskovitz; page 7 inset: iStockphoto/Karen Massier; page 8: Nature Picture Library Ltd./Kim Taylor; page 9 main: Photo Researchers, NY/Tom & Pat Leeson; page 9 inset: ShutterStock, Inc./Tomas Sereda; page 10: Minden Pictures/Wil Meinderts; page 11: Nature Picture Library Ltd./Willem Kolvoort; page 12: Minden Pictures/Ingo Arndt; page 13: iStockphoto/Jason Lugo; page 14: iStockphoto/Arie J. Jager; page 15: Alamy Images/Matt Meadows/Peter Arnold, Inc.; page 16: iStockphoto/Don Wilkie.

Photo research by Ed Kasche; Design by Holly Grundon

Copyright © 2010 by Lefty's Editorial Services
All rights reserved. Published by Scholastic Inc.

SCHOLASTIC, SCIENCE SIGHT WORD READERS, and associated logos are trademarks and/or registered trademarks of Scholastic Inc.

12 11 10 9 8 7 6 5 4   12 13 14 15/0

Printed in the U.S.A.   40

First printing, November 2010

SCHOLASTIC INC.

NEW YORK • TORONTO • LONDON • AUCKLAND
SYDNEY • MEXICO CITY • NEW DELHI • HONG KONG

Fast Fact

A **pond** is a small, calm body of water. It has freshwater like a lake or river.

Let's **visit** a **pond**.

Fast Fact

Hundreds of different animals live in or near a **pond**.

deer

Take a look! You **might spot** all kinds of animals.

raccoon

Fast Fact

Raccoons come to a **pond** to catch fish and drink water.

Let's **visit** a **pond**. You **might spot** a raccoon near the water.

Fast Fact

Frogs are very good swimmers.

You **might** even **spot** a frog.

Let's **visit** a **pond**. You **might spot** a turtle in the water.

beaver

Fast Fact

Beavers cut sticks with their teeth. They use the sticks to build homes.

You **might** even **spot** a beaver.

dragonfly

Fast Fact

This insect can fly 30 miles per hour.

Let's **visit** a **pond**. You **might spot** a dragonfly above the water.

duck

Fast Fact

Ducks can fly, waddle, and swim.

You **might** even **spot** a duck.

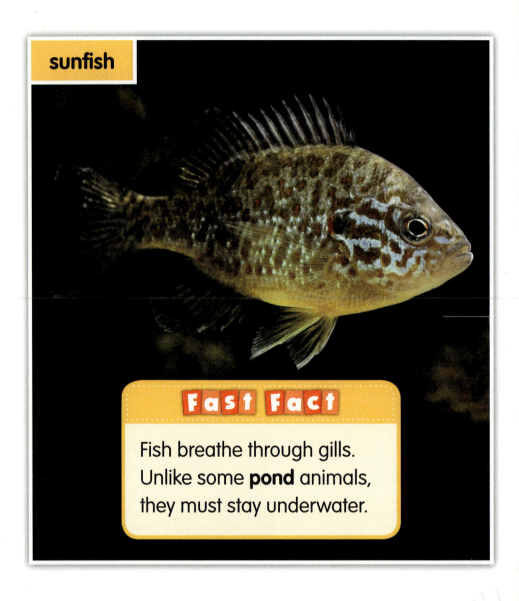

sunfish

Fast Fact

Fish breathe through gills. Unlike some **pond** animals, they must stay underwater.

Let's **visit** a **pond**. You **might spot** fish under the water.

Fast Fact

Snails live on the muddy bottom of a **pond**.

You **might** even **spot** a snail.

water strider

Fast Fact

This insect can walk across the surface of water.

Let's **visit** a **pond**. You **might spot** an insect on the water.

Fast Fact

A **pond** is full of all kinds of life. There are millions of them in the world.

You **might** be surprised by what you **spot**!

Sight Word Review

Point to each sight word. Then read it aloud.

Sight Word Fill-ins

Use one sight word from the box to finish each sentence.

| might | pond |
| spot | visit |

1. Is there a _____ near your home?

2. You might _____ all kinds of animals.

3. You _____ even see a duck.

4. Let's _____ a pond soon!

ANSWERS: 1. pond 2. spot 3. might 4. visit

All About Ponds

Ask a grown-up to read this with you.

A pond is smaller than a lake. A pond is calmer than the ocean with its crashing waves. A pond is also filled with freshwater, unlike the ocean, which is filled with salt water. For all these reasons, ponds make good homes for many different kinds of animals.

muskrat

Because fish breathe through gills, they can live their whole lives underwater. Ponds are perfect homes for freshwater fish, such as bass and sunfish. Many amphibians, such as frogs and salamanders, live there, too. Amphibians spend part of their life on land and part in water. Frogs sit on the sides of ponds looking for bugs to eat. But if a predator comes after them, they can hop into the water and swim away.

Mammals, birds, and insects, such as muskrats, ducks, and mosquitoes, also live in or near ponds. Mosquitoes sometimes lay their eggs in the water. Next time you get bitten by a mosquito, remember that it may have been born in a pond.